I0815378

LET'S CODE!

CODING VARIABLES

BY GEORGE ANTHONY KULZ

CONTENT CONSULTANT
DAVID A. LASH
ASSOCIATE PROFESSOR OF COMPUTER SCIENCE
AURORA UNIVERSITY

Kids Core
An Imprint of Abdo Publishing
abdobooks.com

abdobooks.com

Published by Abdo Publishing, a division of ABDO, PO Box 398166, Minneapolis, Minnesota 55439.

Printed in the United States of America, North Mankato, Minnesota.
102023
012024

Cover Photos: Chesnot/Getty Images Entertainment/Getty Images (game); Shutterstock Images (background)
Interior Photos: Next Mars Media/Shutterstock Images, 4–5; Shutterstock Images, 7, 8, 10, 15, 17 (people, background), 17 (diamond), 17 (hats), 23, 25, 28 (top), 29 (bottom); Nadya Eugene/Shutterstock Images, 12–13, 28 (bottom); Paula Photo/Shutterstock Images, 18, 29 (top); Ground Picture/Shutterstock Images, 20–21; ESB Professional/Shutterstock Images, 24; Chesnot/Getty Images Entertainment/Getty Images, 26

Editor: Katharine Hale
Series Designer: Katharine Hale

Library of Congress Control Number: 2023939665

Publisher's Cataloging-in-Publication Data

Names: Kulz, George Anthony, author.
Title: Coding variables / by George Anthony Kulz
Description: Minneapolis, Minnesota: Abdo Publishing, 2024 | Series: Let's code! | Includes online resources and index.
Identifiers: ISBN 9781098292782 (lib. bdg.) | ISBN 9798384910725 (ebook)
Subjects: LCSH: Coding theory--Juvenile literature. | Computer programming--Juvenile literature. | Computers and children--Juvenile literature.
Classification: DDC 005.1--dc23

CONTENTS

Video games allow players to explore fantasy worlds.

CHAPTER 1

EXPLORING THE FOREST

Liam is playing his favorite mobile game. His **avatar** walks through a forest. He finds a treasure chest. He opens it. There's the key Liam was looking for! He picks it up. It goes into his **inventory**. Then he comes to a house with a locked door.

He uses the key. It disappears from his inventory. A loaf of bread sits on the table. Liam makes his character eat it. His health bar goes up.

Liam is learning about programming in school. He knows video games are written in code. Code is used to write programs. Programs give instructions to computers. The people who write code are called programmers. Someday Liam would like to make his own games. He wonders how his favorite game works. There are so many things going on! The game has to know what his character looks like. It has to know what he is holding. It also needs to know where he is in the game. Liam will learn that each of these things is stored in variables.

Mobile apps and games run on code.

Health bars and scores are common variables in video games.

What Is a Variable?

A variable is a named location in a computer's **memory**. Variables store **data** that can change. In video games, scores go up when players finish tasks. Health bars go down when players are attacked. They go up when players eat food. Items are carried and dropped. Variables store all of this data.

Every variable has a name. This lets programmers know what it is for. It also has an address. The address is where the variable is located in a computer's memory.

Computer Memory

Computers may have billions of memory addresses. It would be hard for programmers to remember what is stored at each one. Variables make things easier. They have names that tell what they are for. A variable named "highScore" is easier to understand than an address like 32343.

Numbers help identify mailboxes. In a similar way, variables help programmers keep track of data.

Each address is given a number. Finally, every variable also has data associated with it. People can think of variables like apartment mailboxes. The name shows who lives there. The number tells where the apartment is located. The mail inside the mailbox is the data inside the variable.

Explore Online

Visit the website below. Does it give any new information about variables that wasn't in Chapter One?

What Is a Variable and How Do Computers Use Them?

abdocorelibrary.com/coding-variables

Buckets can carry different items without changing on the outside. Variables work the same way.

CHAPTER 2

DATA AND NAMES

Variables are used to store data. All programs need data to do their work. Programs use variables when working with data that changes.

A variable is not itself data. It is a place to hold data. In this way, a variable is like a bucket.

The data is inside the bucket. The bucket has a name on the outside. When a program needs the data, it uses the name to find the right bucket. The program can then look at the data inside. When the program needs to change data, it replaces the data in the bucket with new data. The bucket does not change. But the data inside can.

Data Types

All variables have a data type. Data types are used to describe data. They tell the computer what kind of information the data represents. They also tell the computer how to use the data.

Some data types are for whole numbers. A player's score uses this kind of data type.

People often sort their clothes before doing laundry. This is because different colors and fabric types need to be washed differently. In a similar way, data types tell computers and programs how to treat different types of data.

One way this data type can be used is by adding points to a score. When a player defeats an enemy or scores a goal, points are added.

Other data types are for words. Some games let players choose a name for their character. The name might appear in **dialogue**. The variable for the name uses a data type for words.

Simple vs. Complex Data Types

Some data types are simple. They represent one thing. Other data types are complex. They represent collections of simple data items. A birth date can be stored with a simple data type. A list of birth dates can be stored with a complex data type.

Variables in Action

Variables can be used in video games to store data such as a player's name and how much money they have.

It is important for programmers to give their variables meaningful names.

Naming Variables

Variable names are important. They help a program find where data is located. They also help programmers. A programmer can read the code and see what data is being used.

Names should be unique. Each variable has its own data. Having two variables with the same name would be confusing. Because of this, programming languages don't allow different variables within the same **scope** to share a name. Using unique names prevents accidentally changing the wrong data.

Variable names should also be meaningful. A name should show what data the variable holds. A variable named "x" does not say anything about the data inside. However, the name "highScore" shows that the variable is holding the data for the high score.

Further Evidence

Watch the video at the website below. Does it give any new evidence to support Chapter Two?

What Are Variables?

abdocorelibrary.com/coding-variables

Variables help programmers keep track of the data needed for a program.

CHAPTER 3

USING VARIABLES

Variables are useful tools for writing programs. Programming would be difficult without them. A variable is needed whenever data must change in a program. Otherwise, it would be hard to keep track of the program's data.

Bowling with Variables

Suppose a programmer wants to write a bowling game. Lots of data would change. The game would need to keep track of four players' names. These names would change every time different people played the game. The scores for each player would also change. The scores would start at zero. They would

Constants

Some program data does not need to change. For example, the number of sides in a triangle is always three. A variable would not be needed in this case. A constant can be used instead. A constant is like a variable. It has a name to help programmers find it. However, the data stored in it will never change.

A bowling game would need variables to keep track of lots of player data.

increase based on the players' scores for each **frame**. The game would also have to keep track of which pins are up and which are down during each frame.

Real-life bowling uses programming too. Programs keep track of a player's score to tell a machine which pins to set up.

Each of these changing items needs a variable. One might be called "playerOne." It would store a word data type. Player one's

Bowlers earn points by knocking down pins.

name would be stored in this variable. Similar variables would hold the other players' names.

There would be a score for each player as well. These could be "playerOneScore," "playerTwoScore," and so on. Each one would be a whole number variable.

Finally, another variable would hold the status of all the pins for each player. This could be a list. The list tells whether each pin is up or down.

Variables are important for any program with data that changes.

The Importance of Variables

Variables are powerful tools for writing programs. A variable can be used any time data in a program needs to change. Names for variables should tell what data is being stored. This will make reading code easier.

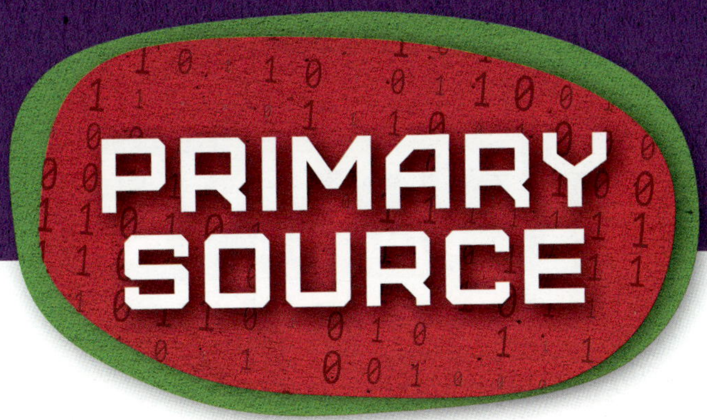

In an article about coding on the Online Journalism Blog, Paul Bradshaw talks about the power of variables:

> Variables can be changed, which is their real power. . . . [Variables] can also be combined: an age (one variable) might be calculated based on a birth date (another variable).

Source: Paul Bradshaw. "Coding for Journalists: 10 Programming Concepts It Helps to Understand." *Online Journalism Blog*, 9 May 2014, onlinejournalismblog.com. Accessed 7 May 2023.

Comparing Texts

Does this quote support the information in this chapter? Or does it give a different perspective? Explain how in a few sentences.

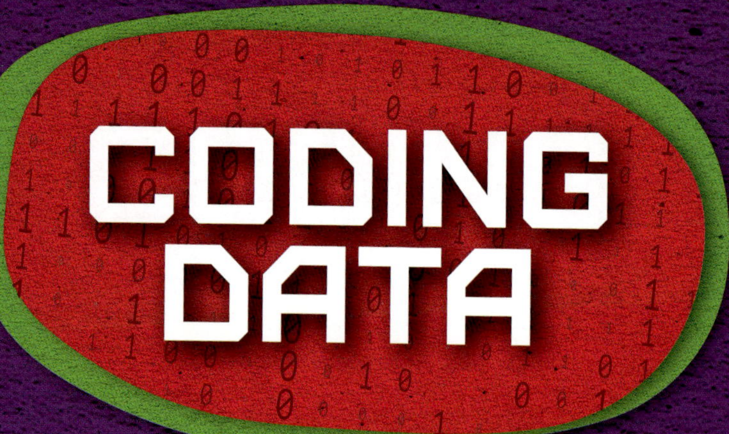

CODING DATA

A variable is a computer memory location that has a name and contains data.

Variables are like buckets. They store data that can change.

Variables should have names that describe the data stored in them.

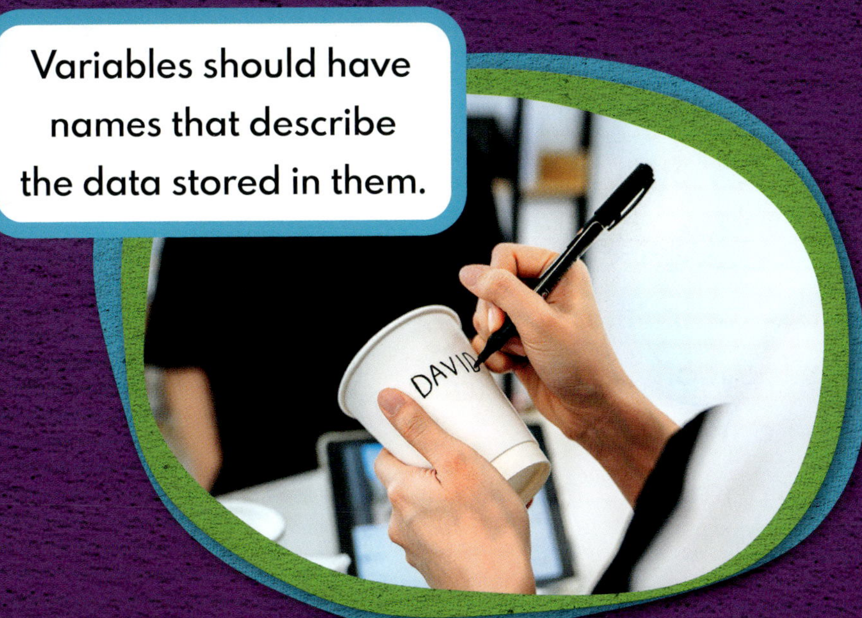

Variables have data types that describe data, tell what type of information is stored in them, and tell how to use them.

Glossary

avatar
in a video game, a character representing and controlled by the player

data
in programming, information that a computer or program can use

dialogue
a written conversation

frame
a round in bowling

inventory
in a video game, the items a player has

memory
in a computer, a place where the computer can keep data and program information and get at it quickly

scope
in programming, a section of code where named items such as variables and functions can be recognized by a program

Online Resources

To learn more about variables, visit our free resource websites below.

Visit **abdocorelibrary.com** or scan this QR code for free Common Core resources for teachers and students, including vetted activities, multimedia, and booklinks, for deeper subject comprehension.

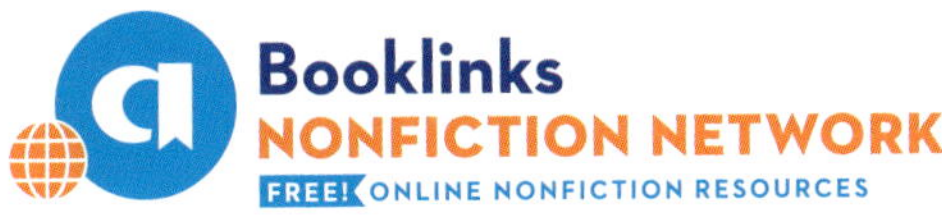

Visit **abdobooklinks.com** or scan this QR code for free additional online weblinks for further learning. These links are routinely monitored and updated to provide the most current information available.

Learn More

Bell, Samantha S. *The Basics of Coding*. Abdo, 2024.

González, Echo Elise. *Organizing Data.* World Book, 2021.

Holmes, Kirsty. *Memory Madness!* Crabtree, 2020.

Index

About the Author

George Anthony Kulz holds a master's degree in computer engineering. He is a member of the Society of Children's Book Writers and Illustrators and has taken courses at the Institute of Children's Literature and the Gotham Writers' Workshop. He writes for children and adults.